CAREERS IN WELDING

WANT A JOB IN A FIELD THAT HAS been around for thousands of years and has dramatically influenced the development of our current world? Welding, the art and science of joining two types of metals, is a career path that offers a wide array of job prospects and good earnings potential.

In America today, welders are in high demand. Many of the currently employed welders are nearing retirement age, leaving openings for newly trained welders. The construction and manufacturing industries, which rely heavily on welders, are busier than ever.

Welding is a career that can be entered into at any age with any kind of background or education. College is not required, though formal training at a technical school is strongly encouraged. However, it is also possible to go straight into the work force through an apprenticeship with little more than a high school diploma or GED and maybe a few welding or shop classes.

Apprenticeships for welding are offered through local unions. Typically, the union pays for three to five years of classroom and on-the-job training designed to prepare the apprentice for certification and promotion to full journeyman status – complete with a significant pay raise.

Anyone can get training without joining a union by enrolling in a welding program at a technical school. Training programs can be completed quickly and are often cost-free. Most technical schools offer financial aid, and some employers will foot the bill to make sure their new welders are properly prepared. Students can start with basic training that will teach them enough to land an entry-level job. There are also more than 80 welding techniques and specialty areas, many of which offer specific certifications. The most successful welders continue to take more classes and obtain more certifications because it is the best way to increase earnings potential.

Starting off in the field of welding, the pay usually ranges between $10 and $15 an hour, but highly trained and experienced welders can end up earning over $60,000 and more a year. The best of the best can make $100,000 plus, annually. The raise in salary can also come from a willingness to work in remote places, such as deep sea welding.

In addition to good wages, a welding career offers a number of attractive features. Welders can work anywhere in the world. The skills and techniques can be applied anywhere, from the bottom of the ocean to outer space. Even in the most common industries, there are opportunities to move around and pursue work that aligns with personal interests. For example, a welder working in an auto factory in chilly Detroit could have a desire to be closer to the beaches where the sun is warm and the fishing is exceptional. The same skills used to build cars can be transferred to building ships. Experienced welders can move into numerous related fields as well, such as research, engineering, teaching, and sales.

Welding offers many opportunities for people who enjoy working with their hands and being in control of the development and potential of their career.

WHAT YOU CAN DO NOW

THERE ARE MANY THINGS YOU CAN do to prepare for the career while still in high school. Because being a welder does not require a college education, it is even possible to go straight into a job or apprenticeship as a welder right after graduation.

Take every shop course you can in welding and metal fabrication. Some high school classes use a virtual stimulation program that will show you what it feels like to be performing welds. If you are lucky, you will get real hands-on experience with a good instructor. Either way, a welding class will teach you about the different types of welding.

Do not assume shop is the only important class. Good math skills are essential for success in any welding job. This goes beyond simple addition, subtraction, multiplication and division. You also have to be good at problem solving and know basic geometry. Science is important, too. After all, welding is a kind of science. You need a basic understanding of how and why welding actually works before you can do it.

Talk to working welders. You can find them at local welding association groups, companies, or union halls. The goal is to find someone who would be willing to let you job shadow for at least a day. Just watching a welder in action can give you a better feel for the conditions you can expect to be working in and what the job is like day to day. Be sure to ask questions. Find out how the welder got started and ask for advice on how you should prepare.

Tagging along with a professional will also help you line up personal contacts, which can be important once you are in the job market. If you develop a good relationship with the people you go on jobs with, they may help in getting you hired as an apprentice after you graduate.

The best welders are well-rounded people. You will soon learn that in just about any welding job you need to work with other people. Join clubs. Be part of a team. Participate in debates to learn how to talk a problem out. Try new things and experience something completely different. It will make you a better person, and ultimately, more successful in this, or any other career.

Look for chances to get real work experience with welding. Try to work part time after school or during the summer in a body shop or tractor repair shop.

HISTORY OF THE CAREER

ROUGHLY 5,500 YEARS AGO, A smart Bronze Age worker came up with a great idea. What if he heated the bronze alloys he was working with until they melted, and then pounded them together with a hammer to create the shape or form that he wanted? It worked, and industrious people have been welding ever since.

Examples of early welding techniques are dotted throughout ancient history. Around 3000 BC, the Sumerians in Mesopotamia (modern Iraq) were making swords that were joined by hard soldering. About the same time, the Egyptians heated iron ore in charcoal fires to reduce it to sponge iron so that pieces could be hammered together (solid phase welding). Smelting, extracting metal from its ore by heating and melting was common by 1500 BC. A wall painting at the tomb of Vizier Rekhmire at Thebes depicted a brazing operation in 1475 BC. Brazing is uniting metal objects at high temperatures by applying solder.

During the Middle Ages, forge welding was advanced by blacksmiths who pounded heated different metals together repeatedly until bonding occurred. Renaissance craftsman were skilled in the process, and the welding methods continued to advance during the following centuries.

Welding took a dramatic turn in development once electricity became a part of the innovation in the 1800s. The electrical arc was discovered in 1801 by Sir Humphrey Davy. Vasily Vladimirovich Petrov used the electric arc in 1802, building the world's largest and most powerful voltaic pile at the time, which consisted of around 4200 copper and zinc disks. In his work published in 1803, Petrov proposed the usage of the electric arc in welding, having managed to perform simple experimental welding. It was not until the 1880s that the technology became developed with the aim of industrial usage.

Acetylene was discovered in 1863 by Edmund Davy, but its use was not very practical for welding until around 1900, when a suitable blowtorch was developed. At first, oxy-fuel welding was one of the more popular welding methods due to its portability and relatively low cost. As the 20th century progressed, however, it became less popular for industrial applications. It was replaced for the most part by arc welding, as metal coverings (known as flux) for the electrode were used to stabilize the arc and shield the base material from impurities.

The Industrial Revolution of the late 19th century created a boom in construction and manufacturing in America. The result for welding was rapid growth and innovation in the field. World War I also pushed the growth of the welding industry, as each country attempted to determine which of the several new welding processes would be best. After the war ended, major advances in welding technology continued. Stud welding was developed in the early 1930s. It was a form of spot welding that welded a bolt or specially formed nut onto another metal part.

Underwater arc welding was also pioneered in the 1930s. This technique was originally developed by the Linde - Union Carbide Company, but it was successfully implemented by Soviet engineer Konstantin Khrenov. Its development has been further advanced and is still very popular and used throughout the world today.

Gas tungsten arc welding, after decades of development, was finally perfected in 1941, and gas metal arc welding soon followed in 1948, allowed for fast welding of non-ferrous materials but required expensive shielding glasses.

Increasing welding speeds has been the goal for many inventors since the 1950s. The flux-cored arc welding process first appeared in 1957. It used self-shielded wire electrodes with automatic equipment, resulting in hugely increased welding speeds. In 1958, electron beam welding emerged as the best way to use a highly concentrated heat source in the process. It laid the foundation for the later development of laser beam welding, which is now very useful in high speed automated welding. In 1976, the first automotive production application for laser welds began at General Motors in Dayton, Ohio. By 2001, CO_2 lasers were being used to weld polymers. Using silicon carbides embedded in the surfaces of the polymer, the laser is capable of melting the material, leaving a near invisible joint line. In 2013, the joining of aluminum and low-carbon steel was achieved with the laser, using a lap joint.

Research and development in the field of welding is ongoing. Today, nearly 25 percent of welding is automated – a trend that is predicted to grow by roughly 20 percent over the next few years. As developing countries build infrastructure and train more workers in the field, international competition for welding grows, and research is continuing to keep the industry on the cutting edge of efficiency and safety.

From melting bronze down and beating it with a rudimentary hammer, to using high-tech laser technology, welding has played a crucial role in the development of infrastructure and society. Its methods have been continually advanced because of the importance of welding in almost anything that needs to be constructed or manufactured.

WHERE YOU WILL WORK

WELDERS ARE IN HIGH DEMAND across all industries that require any kind of construction, or which work with metal. Any kind of metal, no matter the size or the type, can be shaped through the process of welding. Because of the wide range of applications for welding, welders can be found working just about everywhere.

The majority of welders work in some type of manufacturing. Consider all of the everyday things that are in high demand and require metal, from cars, trucks, motorcycles, rail cars, ships, and aircrafts, to rockets and space stations. The factories that make all of the parts for all of these products use a combination of trained welders and machines to handle all of the metal that needs to be shaped and sized to fit the product. Some welders in factories work hands on (or directly) with the metal, while others are trained to operate machinery, like enormous welding arms that make the welds.

Many welders also work in construction. The construction industry is enormous and vital to all countries and cities, especially those that are developing quickly. Skyscrapers, highways, bridges, pipelines, offshore oil platforms, wind turbines, and solar panels all require the help of welders. The need for welders in any kind of construction that uses metal means that you could be installing and maintaining anything from boilers to antipollution systems for industrial, commercial, or residential facilities.

Not all welders do on-site welding in factories, shipyards, or large construction sites. Some work in schools and offices, using their background in welding to teach or work in sales. Others advance to management positions, or go back to school to become engineers.

Work environments vary. Conditions can be difficult or even dangerous. Some welders can be found working on construction beams of high rises, sitting out in the open air many stories up, while doing their work. Some work deep underwater or underground. Others have to perform their work in tiny, cramped spaces. Those who prefer to stay in familiar conditions usually prefer working in manufacturing, where tasks are repetitive and the conditions never change. However, there are certainly many more interesting and challenging positions – out at sea, deep underwater, on the beam of a high rise, or even higher, in outer space!

THE WORK YOU WILL DO

WHETHER YOU ARE ON THE OCEAN floor, in a warehouse, or in outer space, the basic work a welder does every day is essentially the same. Welding is the process of permanently joining metal parts. The techniques used will vary, but every task comes down to taking pieces of metal and turning them into something else. By far, the most popular technique is arc welding.

Arc welding is the foundation of all basic training provided by technical schools. While there are more than 80 different types of welding methods, arc welding is the most commonly used today. Arc welding uses an electrical current to create the necessary heat for bonding metals together quickly and effectively. The electrical current that is used in this process also allows the metals to cool quickly and makes a solid bond between the metals possible.

In arc welding, an electrode, which is also the filler, is continuously fed through the nozzle of an arc torch. When you activate the torch, several things happen. As the electrode begins feeding through the nozzle, a direct current is generated that creates an arc when it comes into contact with the electrode. Shielding gases are simultaneously released around the nozzle to protect the weld from atmospheric gases that could degrade its quality. Once the arc has been created, the welder can move it around the weld joint and create the weld.

There are two basic kinds of arc welding. The most common is Tungsten Inert Gas (TIG). This is an advanced type of arc welding that is typically used with materials like stainless steel and aluminum. During this process, a welding rod is used in combination with an electric torch. The electric torch simultaneously melts the rod and the piece of metal.

Metal Inert Gas (MIG) welding is another common yet advanced arc welding technique. MIG uses a feeder to continuously distribute a wire that joins long stretches of metal. The wire allows the welder to continuously weld without having to stop and replace a rod.

The work you do as a welder is both wide-ranging and simple. There are many techniques and many places you can perform welding, but at its heart, welding is simply the joining of two molten metals. No matter if you are using a robot or TIG, the basic principle remains the same. Beyond the basics, however, there is a wide variety of welding jobs. Some are entry level, while others require advanced training and experience. Some are done in ordinary factories, while others are done in extraordinary environments. Here are some of the most common types of welding jobs.

Welding Technicians

Welding technicians conduct experiments and tests to evaluate new or improved welding equipment, techniques, procedures, and materials. They also inspect welded joints to make sure that company standards and national code requirements are being met. There is a considerable amount of paperwork involved in this type of job. Welding technicians must record the results of all inspections and tests. They must also prepare and submit technical reports to engineering and management personnel. The data must be accurate, as it will be used to determine the outcome of research and development projects, and in preventive maintenance investigations. Welding technicians also conduct certification tests for welders trying to meet national code requirements.

There is a considerable amount of paperwork involved in this type of job. Welding technicians must record the results of all inspections and tests. They must also prepare and submit technical reports to engineering and management personnel. The data must be absolutely accurate, as it will be used to determine the outcome of research and development projects, and in preventive maintenance investigations. Welding technicians also conduct certification tests for welders trying to meet national code requirements.

Welding technicians work closely with engineers, assisting with the testing of materials, metals, and alloys. They might recommend that new developments and applications be adopted for a particular project. They judge whether improved technologies and techniques recommended by the engineers are carried out correctly, and whether efficiency has improved as a result.

Some welding technicians do work that is more hands-on. For example, they might help set up manual welding equipment, troubleshoot manufacturing operations, program and monitor automated welding machinery, and oversee factory maintenance.

Welder-Fitters

Welder-fitters lay out, position, and secure parts, and assemble them according to the specifications on blueprints. To do this, they use mathematical tools such as straightedges, combination squares, calipers, and rulers. The welder-fitter typically has more expertise in blueprint reading and assembly than most welders do. In general, the welder-fitter may have a stronger background in math and often works as a supervisor or engineer.

Sometimes, especially on larger projects, the work will be divvied up to two separate people – a fitter and a welder. The fitter reads the blueprints, marks the weld locations, fits up the pieces, and tacks things in place. Once the components are tacked together, the fitter will decide whether it will meet specifications, and then send it on to a welder to finish welding the joints together. However, it is more common for the same person to do both tasks.

Welding Inspectors

Welding inspectors are highly trained to perform tests on welds to ensure the quality and safety of welded work in various structures. Determining whether weld joints can handle stress is essential in making sure that buildings and infrastructure are safe for public use. Inspectors often work on hi-rises and other structures high off the ground. Certified welding inspectors are experts in every aspect of welding. In addition to inspecting, they have knowledge of all welding techniques, and are able to weld and test metal samples.

In order to conduct some of the necessary welding tests, such as checking for cracks and testing welding strength, an inspector may need to use special lights and magnifying glasses. These are also utilized to determine whether a welding joint has any cold welds or undercuts. Welding inspectors also conduct ultrasound testing. This is an important test that can discover any defects and cracks that can jeopardize the safety of a structure.

Welding Engineers

Welding engineers develop welding techniques, test new fabrication processes and procedures, and modify and improve welding equipment. In addition to engineering principles, they have thorough knowledge of all kinds of metal fabrication, production specifications, and the characteristics of various metals and metal alloys.

On a typical day, a welding engineer would conduct research and development projects, prepare and write technical reports, and test welds to show they conform with national code requirements.

Engineering is a team activity. Welding engineers work closely with various project team members. They often direct and coordinate technical personnel from government agencies who perform inspections to ensure compliance with established welding procedures, restrictions, and standards. They also test welding personnel for certification. They get together with clients to discuss ideas, exchange information, and offer technical advice about welding.

The work welding engineers do is important and constantly changing. They must stay on top of the latest developments in the welding field to know whether they can possibly be applied to current welding problems or production processes.

Aerospace Welders

Aerospace welders do the same basic work as all welders – they operate equipment that uses heat to join metal parts together. In the aerospace industry, this means they work on manufacturing parts and vehicles, from airplanes to space shuttles. This may require working with metals like carbon steel, stainless steel, and aluminum.

Most aerospace welders work in large factories that produce airplanes, but a few get to work in outer space, repairing the space station.

One of the techniques most commonly used in outer space is Gas Tungsten Arc Welding (GTAW), which is a relatively low-heat method that uses a non-consumable tungsten electrode. Its low-heat characteristic reduces distortion in thin metals, which are often used in space.

Submerged Welders

Submerged welders work underwater. Their work environment is the most dangerous, exciting, and high paying of all the welding specialties. Because of the dangers of shock, explosion, and poisoning, only professionals with both diving and welding certifications perform underwater welding.

Like all welding, submerged welding is still a process whereby metals are melted together to either repair or create a new structure – but it is done underwater. Submerged welders work on offshore oil platforms, pipelines, ships, and other underwater structures.

The most common technique used is hyperbaric welding, in which a structure is built around the weld and a pressurized, dry environment is created. The "hyperbaric habitat" is essentially a large box that allows the welder to have a dry environment in which to work. The hyperbaric habitat used on an underwater welding job can be so small there is just room enough for the welder to insert two hands and manipulate welding tools, or so large that several welders can climb in and work multiple welds simultaneously. Hyperbaric welding can also take place in the water itself. Even though the method is the same, when welding is done in a wet environment, it is called "underwater welding."

STORIES OF WELDERS

I Operate Robots

"I work at a power plant, doing orbital welding using robotics. Orbital welding is a specialized type of welding where the arc is rotated mechanically 360° around a static workpiece, such as a pipe, in a continuous process. Like any other welder, I wear full protective gear, but I'm sitting at a computer monitor, controlling the robot that does the welding. playing a video game. The only downside is the work schedule. I work all night on a 12-hour shift.

This probably sounds like a career for nerds, but it's not. I was never good at math and computers didn't interest me. My motivation was earning money. I wanted to earn a good living and my high school counselor advised me to go to a technical college and see what clicked for me. I took the advice, and the first time I welded, I knew it was what I wanted to do. After learning how to do fabrication with hand welding, I continued my training with orbital welding and robotics.

I love my job. The pay is great and I know I can work anywhere in the world. I never expected a career working with computers, but that is where welding is headed. The need for welders is going up, despite the increased use of automation. That is because welders are still needed to adjust welds, operate the welding machinery, and make programming adjustments.

Welding is a field where you can control your own destiny. The biggest advantage you can give yourself is a good foundation in welding technologies."

I Am an Industrial Artist

"I was a fabricator for 15 years before deciding to start a little shop of my own doing auto body work. Like most welders, I have a passion for what I do. I'm not happy unless I'm welding, even in my dreams! It was slow going at first in my startup business, so I had plenty of time on my hands. I started playing around with recycled auto parts and other bits and pieces of scrap metal. My first sculpture was a dragon with bulging round eyes and a comical grin. It was cute and funny, but I never thought it was worth anything – until a friend came by and asked if he could buy it. Suddenly, I wasn't a welder anymore, I was an artist!

It took a while for my welding skills to evolve fully into an art form. Taking a few art classes at a community college helped me loosen up and view my work with a different eye. Now I see my cutting torch as a brush, and steel is my medium. I no longer look at two pieces of metal figuring out how I can join them to make them functional. The act of heating metal to its melting point changes the physical characteristics of the metal, even changing its shape and color. I've learned to use these distortions to enhance the artistic beauty of my sculptures.

Today, I make my living entirely from welding art. Most are whimsical pieces for the home and garden, sculptures of anything from crickets for the hearth, to a man on horseback. I still like to combine function and design. To that end I create wine racks, occasional tables, lamps, and anything else that pops into my head. No longer am I limited to using car parts. My shop is filled with old drain pipes, sewer rings, industrial gears, springs, and flatware. Anything with an unusual shape catches my eye and I drag it home. Sooner or later, it will end up in a sculpture, representing an ear, eyelashes, or an anatomical part that doesn't exist in real life.

My advice for the aspiring metal sculptor is to practice, practice, practice. The importance of good craft skills shouldn't be underestimated. I have nearly 20 years of welding experience, which allows me to apply the best technique to get the results I want. It's not easy and takes a lot of practice. Also, learn to draw. Remember, there are two parts to this work. One is welding, the other is art. Drawing and sculpting classes will help you immensely. There are a few art and design schools that offer classes in welding and industrial art, but any art classes will be beneficial. Practice drawing. It helps you visualize and define your sculpting ideas.

Welding as an art is a skill that can be learned and enjoyed. It can provide a nice income if you want it to. There are pieces of welding art that have sold for hundreds of thousands of dollars. I'm not at that level, but I earn as much as I ever did as a fabricator. And this is a lot more fun!"

I Weld in the Ocean

"I always wanted to travel and have new experiences in life. I didn't know how I was going to do that until I read an article on underwater welding. Welding was not something I had ever considered, but it turns out I am good at it.

After getting my 'topside' welding certificate through a technical college, I worked as a fabricator to get some experience and improve my skills. At the same time, I enrolled in a commercial diving school. On land, I was always having to prove myself because I was a woman in a male-dominated field. In the water, it's completely different. My first time welding in

the water was the most exciting thing I had ever experienced. It's another world down there. No one is down there telling me what to do or questioning my ability. It turns out employers like women divers because they naturally pay attention to detail, are more patient, and are meticulous in their work.

My job is so cool. I have the freedom to go anywhere in the world – and I do. I never know where I'll be one week to the next. I could be making an emergency repair on a cruise ship docked in Dubai today, and spot weld a pipeline in Curacao on Monday. The money is really good so I can be independent and I like that, too.

There are almost infinite career choices a welder can make after graduation. Underwater welding is just one. Jobs are opening up for American welders all over the world. Manufacturers, construction companies and especially oil companies are always looking for new talent. Welders on offshore oil rigs or those who operate in uncomfortable physical conditions or underwater, stand to earn at least twice what technicians would earn topside."

PERSONAL QUALIFICATIONS

WELDING IS A PHYSICAL JOB THAT requires good eyesight and excellent manual dexterity. A great welder has the exceptional hand-eye coordination necessary to perform the complicated physical maneuvers.

Welders sometimes work solo and sometimes in collaboration with other workers. If you choose to work in an environment with other welders (which is more likely), it is important to be a good team player. This often means taking direction from other people. On a typical job site, there will be supervisors and engineers whose job it is to make sure the correct procedures, equipment, and techniques are used. You may think you are going about a weld in the best way, but if a supervisor tells you to use a different method, you must assume there is a good reason and do as you are instructed.

Whether you are working by yourself or on a team, you must be able to concentrate for long periods of time. When planning, executing, and inspecting a weld, it is extremely important that you stay focused and alert, as the consequences of poor quality work can be enormous. Mistakes often result from being tired. Welding is strenuous work, so it is important that you take care of yourself and stay in good physical condition.

Because there are over 80 welding techniques and the field is always advancing, the most successful welders are those who are eager and willing to learn something new. Taking more classes and improving your knowledge will pay off in more job opportunities and higher wages.

ATTRACTIVE FEATURES

WELDING IS CAREER WITH MANY attractive features. There is good earnings potential, starting at about $15 an hour, as a welder straight out of high school without formal training. That pay can rise to over $100,000 a year for welders who are highly trained and experienced, and who are willing to work in remote locations. Welders are encouraged to continue training and become qualified to do more types of welding, something that generally results in higher pay. It is common for large companies and unions to pay for that training – a big plus!

Welding also offers the opportunity to see the world. Construction and manufacturing projects that require welders are everywhere – right where you live, but also in other states or even overseas. You can even work in the sea. With some specialized training and a diving certificate, you could become a deep-sea welder, installing and maintaining pipelines, oil rigs, etc. If being on a different continent or deep underwater is not remote enough for you, there is one more frontier for welders: space. Believe it or not, there is a demand for welders in outer space, too.

Perhaps the most attractive aspect of this career is the ease of entry. Anyone, at any point in   life, can get into welding even with little or no welding background. A high school student, upon graduation, can go straight into an apprenticeship and then into the job market. Anyone looking for a career change will find it is easy to get certified at a technical school within a year or less. Finding a job quickly at that point is all but guaranteed because there is so much demand for qualified welders.

UNATTRACTIVE ASPECTS

A DOWNSIDE TO WELDING WORK IS the risk involved. Using gas or lasers to melt metals and reshape them means being in close contact with extremely high levels of heat and light. Welds are also often made in difficult or challenging places, with little room to move around. Even small errors can lead to disastrous consequences. In short, welding can be dangerous, not just for the welder, but for others as well. If the joints created by welds are not done properly, the public could be in danger of a building or bridge collapsing.

Welding can also be physically strenuous. Those working on oil rigs, for example, typically work 12-hour shifts. That kind of schedule can be exhausting. Welders need to be in great shape to have the physical and mental stamina necessary to consistently produce good work.

Some welding jobs are repetitive and boring. Manufacturing jobs, for example, often involve doing the same welds over and over again. It is certainly possible to work at more interesting jobs in more appealing surroundings, but those jobs are reserved for those with additional training, certifications, and experience in the field.

EDUCATION AND TRAINING

EXTENSIVE SCHOOLING IS NOT NECESSARILY a requirement for becoming a welder. For example, you could go to a union hall with nothing more than a high school diploma or GED. That would be enough to qualify for entry into a union apprenticeship, where the training is paid for and the apprentice is also paid while learning on the job.

It is also possible to get an entry-level nonunion welding job with just a few weeks of welding training in a high school shop class. That would limit the beginner to low-level positions with no prospects of pay raises until additional training is obtained.

Rather than jump into this career unprepared, it is far better to get some training first. Although some employers provide on-the-job training, they much prefer to hire workers who already have experience or formal training.

There is quite a range of training programs available, from a few weeks at a technical school to four-year degrees from universities such as LeTourneau University, Ohio state, Ferris State, and the Colorado School of Mines. The US Armed Forces operate welding schools as well.

The most popular choice is a technical school because it is the easiest to get into. You can apply to technical school after you have graduated from high school or have your GED. There is no previous knowledge in welding required. Technical schools typically offer financial aid and many employers will even pay for your training.

Basic training programs include courses such as blueprint reading, shop mathematics, and mechanical drawing, as well as hands-on experience in MIG welding. More advanced training is generally separated into modules, any of which can be used to qualify for additional certifications. The most common modules are:

Welding Metallurgy covers the concepts and fundamentals of atomic structure, grain structure, heat flow, phase transformations, welding metallurgy, and the "weldability" of ferrous and non-ferrous commercial alloys.

Joining and Cutting Processes introduces the basics and principles of major joining and cutting processes. Classroom discussion covers the advantages, disadvantages, equipment, consumables, techniques and variables for each process. Students learn how to set up and operate the welding and cutting equipment for specific applications.

Robotic Welding is an increasingly popular module. Here, students learn the concepts and fundamentals of robotic welding, from design to implementation. A background or at least some understanding of computer systems is very helpful in learning how to program and operate robots for welding.

Weld Quality and Inspection prepares students to become certified welding inspectors. Classroom study covers the concepts and fundamentals of welding codes, specifications, and safety guidelines. There is considerable laboratory work where students learn weld quality and inspection methods. They also learn how to set up and operate the instruments and equipment used to identify weld defects.

Laser Welding Technology is another popular module because the methodology is used in so many applications today. The coursework usually covers basic optics, laser welding systems, and optimization of laser welding equipment.

When considering the choices, it is very important that the program be accredited. That is because many cities and states require welders to have a license, and many employers require certification. In any case, your employment application will require evidence of having passed tests meeting the standards set by the American Welding Society Standard Code. The standards include a stipulation that students have attended an accredited program. The largest national accrediting organizations are the Accrediting Commission of Career Schools and Colleges of Technology (ACCSCT), and the Council on Occupational Education (COE). Both are recognized by the US Secretary of Education.

Apprenticeships

A union apprenticeship typically includes three to five years of training, paid for by the union. Each year of training will consist of several thousand work hours and a few hundred classroom hours. Upon completion, an apprentice will enroll in the welding certification program that is sponsored by the union through collaboration with the American Welding Society. In some cases, it is possible to pursue welding certification simultaneously during the apprenticeship program.

Once certified, an apprentice can choose to either continue the education in learning more specialties, or apply to jobs through the union hall that only require basic knowledge. Union workers usually continue to obtain more education and experience. Depending upon the jobs they do, they will be required to re-test for certification anywhere from every three to nine years.

Certifications

Some welding positions require general certifications in welding, while others require certifications in specific skills, such as inspection or robotic welding. The American Welding Society certification courses are offered at many welding schools. Some employers have developed their own internal certification tests.

Welder certification (also known as welder qualification) is based on specially designed tests to determine a welder's ability to create welds of acceptable quality following a well-defined welding procedure. Most certifications expire after a certain time limit, and have different requirements for renewal or extension of the certification.

EARNINGS

WELDING CAN BE A WELL PAID career path. Without any training, beginner welders can expect to start off making about $15 an hour. However, with some time and experience, that modest income can double or triple. In general, pay rates vary according to the size of the employer, industry, geographic location, and the welder's personal expertise.

Small companies may only have one to five welders, while medium companies have eight to 12, and large companies employ 20 or more. Generally, the larger the employer, the more they can afford to pay. Plus, large companies are more likely to be union shops, which typically pay higher wages.

The two industries employing the largest number of welders are manufacturing and construction. Overall, pay rates are similar for both industries. However, there are differences among specific types of work. For example, a welder working for a shipbuilder will likely earn about 15 percent more than a welder doing the same job for a car manufacturer. There are other industries that offer higher pay, though there may be fewer jobs available. These industries include electric power companies, airlines, natural gas distributors, paper mills, and metal ore mining companies.

Geography plays a role in determining salaries for several reasons. First, some areas have more union shops than others do. Also, the cost of living of the area usually influences salary levels across the board. Some of the highest paid welders live in Alaska, Hawaii, and California. Higher paying industries tend to be located in specific areas rather than all across the country. An obvious example would be shipbuilding, which can be found in Louisiana and Mississippi, but not Utah or New Mexico.

Nothing influences a welder's income level more than personal expertise. The more trained and talented you are at welding, the more you can earn. For example, steel is an easy metal to weld. Anyone with rudimentary MIG skills can work on steel and earn about $30,000 a year as a tank welder. It is much more difficult to work with aluminum. A welder with the expertise to work with this metal can earn $70,000 or more in the aeronautics industry, building airplanes. Simply put, if you are able to do the work that fewer people can do, you will make more money.

The pay for specialty welding is much higher than basic welding. There are certifications for all kinds of specialties. Some are based on the type of materials, such as stainless steel. Others are based on the type of work, such as programming robotic welding.

One of the highest paying specialties is underwater welding. Salaries for underwater welders cover a wide range, from $60,000 to $150,000 or more annually for full-time work. The majority of underwater welders are paid on a per-project basis, and income is naturally highest where there is consistent work available. Many underwater welders only get sporadic calls to use their special skills. How much an underwater welder works typically depends on the geographic location. Most jobs of this kind are located in the Gulf of Mexico where there are offshore oil rigs. Therefore, welders living in Gulf states such as Texas and Louisiana are finding increasing opportunities to work full-time while those living in New York may only get occasional calls. Other factors that affect pay rates include depth, dive method, and diving environment.

Union Versus Non-union

About 20 percent of all welders are union members. Joining a union can have its advantages – paid training, steady work, and great benefits. In general, union welders earn a higher salary than non-union workers, although this varies greatly across different specialties. The average union welder's annual salary is about $90,000 a year, which is considerably higher than the average pay for non-union welding workers. However, there are union-shop jobs that pay less than non-union shop jobs, so make sure to do your research before joining a union.

As an apprentice, a union welder might earn $15 to $18 an hour, which is $25,000 to $37,000 annually. It is not unusual to stay at the same income level for five years without a raise. However, that quickly changes once the apprentice completes training, gets certified, and becomes a journeyman. New journeymen make roughly $40,000 to $50,000 a year. With each new certification and more experience comes more pay.

OPPORTUNITIES

THE JOB OUTLOOK FOR WELDERS IS bright. Experts predict about 15 percent growth for welders over the coming decade. That amounts to an additional 50,000 welding positions. Many jobs are being opened up by Baby Boomers who are entering retirement. There is also tremendous growth in the aeronautics industry, where welders are needed to build missiles and aircraft for the US military as well as commercial aircraft. Rebuilding infrastructure is another source of job growth. It is impossible to rebuild the nation's aging bridges, highways and buildings without welders.

Considering this healthy forecast, you can afford to think about where you want your career to go once you have a foothold. There is a wide range of demand for welders, and many interesting opportunities within the field. Want to go deep sea diving? Want to go to outer space? Want to travel the world? Anywhere anything is being built, welders are required – and that means just about anywhere you can imagine. You can take your work to pipelines deep underground or underwater, to oil rigs far out at sea, or to solar panel fields. You can stay close to home, work inside, keep normal hours, and enjoy a steady paycheck.

The American Welding Society (AAS) reports that there is particular need for underwater welders. The demand has been increasing due to multiple factors, including deep-sea oil and gas development, and marine infrastructure repair and development. It is important to note that the most experienced and skilled underwater welders are the ones employers call first. Breaking into the field requires patience and determination. However, it is possible to greatly improve job prospects by choosing an accredited underwater welding school, ensuring the program provides a maximum of hands-on training and experience, and pursuing the necessary certifications. In addition, entry-level beginners should be flexible about location. Those who are willing to go where the work is, are much more likely to secure a position.

It is often possible to incorporate special interests or hobbies into a welding career. For example, if you have always loved cars and auto racing, it is possible to work as a welder on a team for a race car driver. Before any NASCAR hits the road, almost 1,000 man hours are spent on welding and fabrication for each car, as hundreds of parts have to be hand cut, welded, and machined.

Even a passion for art and sculpture can be pursued through welding. This is not as unusual as you may think. There are numerous self-employed welders who work with artists who need help on installations that involve welding. Some artists will want to do the welding themselves, but they may require an expert to help teach them and be there in case the artist needs any help.

The highest demand for welders will remain in the manufacturing sector. However, because basic welding skills are the same across industries, welders can easily shift from one industry to another. For example, welders who previously worked in automotive manufacturing might choose to find work in the oil and gas industry.

Whatever your preferences, you will have more choices if you are trained in the latest welding techniques and welding devices. Advanced training leads to greater job security and higher pay. Welders who are not trained in the latest technologies and equipment will have a harder time finding placement and might have to relocate to find better employment prospects.

GETTING STARTED

DO NOT WAIT UNTIL YOU GRADUATE from technical school to start laying the groundwork for your job search. Teachers can be excellent resources since many of them are active welders themselves. Their connections can be your connections if you do well in class and show that you have the potential to be a great employee. Some vocational schools and community colleges work directly with local employers to find graduates from welding programs. Check in with your school's placement program regularly to learn about any new openings for beginners.

Look where the jobs are most plentiful. The biggest employers are in construction and manufacturing industries. There are also geographical pockets where welders are in demand, such as the Gulf Coast states, where there are shipbuilders, oil refineries, and offshore rigs.

You can apply directly to any company that needs welders. Most of these companies will have standard applications. You will need to do more than fill out an application if you want to edge out the competition. Prepare a résumé that clearly notes all of your certifications and related experience. If you do not feel comfortable writing a presentable résumé yourself, there are plenty of professional résumé writers who can help. Get letters of recommendation from teachers, employers, and experienced welders you have met. During the interview, make sure you leave the impression that you are willing to take on any task, and are eager to learn everything you can about your new career.

Get involved in the welding community by joining professional associations or societies specifically for welders, such as the American Welding Society (AAS). Joining a professional organization is a good way to make contacts that could lead to job opportunities. Wherever you go, let people know you are willing to travel, relocate, work long or irregular hours, start at the bottom, and do the dirty work needed to prove yourself. Do not be afraid to ask the more seasoned welders you meet for tips and advice.

Get on the Internet. Join welder groups on professional and social networking websites. Employers look to organizations such as the AAS and networking groups to find welders for their job openings. There are numerous large databases online that offer access to employers specifically seeking welders. Make a daily ritual of checking the general job boards, too, such as Indeed and Monster Jobs.

Get comfortable with social media. Join LinkedIn, set up your profile, and start searching for professionals in your field to add to your network. Join groups on Facebook that are dedicated to welders, both newbies and pros. And do not forget YouTube. There are instructional videos that will teach you how to get a welding job as well as some job announcement videos from major employers.

Get experience any way you can. In welding, education is just the beginning. Experience is key. Look for an entry-level job doing MIG. It is hard, dirty work, but it is the kind of job where you learn the most when you are starting. Take more courses whenever possible, particularly in metallurgy and some of the finer details of welding. The more certifications you have, the more employable you become.

Get a union job. The first step is to contact the welding union in your area to find out what qualifications you will need prior to acceptance into the local chapter. The second step is to complete an apprenticeship program. Apprenticeships usually last three to five years and offer paid training. After five years, you will move from an apprentice to a journeyman welder, with a significant raise in pay.

ASSOCIATIONS

■ **American Welding Society**
http://www.aws.org

■ **United Association**
http://ua.org

■ **National Certified Pipe Welding Bureau**
http://www.mcaa.org/ncpwb

■ **Fabricators & Manufacturers Association, International**
http://fmanet.org

■ **International Association of Bridge, Structural, Ornamental, and Reinforcing Iron Workers of America**
www.ironworkers.org

PERIODICALS

■ **The Fabricator**
www.thefabricator.com

■ **Miller Welds**
http://www.millerwelds.com

WEBSITES

■ **Careers in Welding**
http://www.careersinwelding.com

■ **Go Welding**
http://www.gowelding.org

■ **Welding Web**
http://www.weldingweb.com

■ **WeldingTipsandTricks**
www.weldingtipsandtricks.com

■ **Arc-Zone**
http://www.arc-zone.com

Copyright 2015

Institute For Career Research

Careers Internet Database Website

   www.careers-internet.org

Careers Reports on Amazon

   www.amazon.com/Institute-For-Career-Research/e/B007DO4Y9E

For information please email service@careers-internet.org